Barns of ARKANSAS

Electric Cooperatives of Arkansas

Your Local Energy Partners

www.ecark.org

Your Touchstone Energy® Cooperatives

Gus, the goose, and his friends spent a lot of time in this pasture. Chester and Joyce Woodruff, of **Howard County**, snapped this photo of Gus and friends in front of their barns, tool shed, and equipment sheds. All of these buildings were destroyed by a tornado in December 1999. *(Submitted by Deb Wallis, Nashville FFA Chapter)*

THE
DONNING COMPANY
PUBLISHERS

Barns of ARKANSAS

by **Angela B. Collins,** Arkansas FFA State Executive Secretary

Compiled as a special project of the Arkansas FFA Foundation and sponsored by the Electric Cooperatives of Arkansas

Endleaf: Wayne and Kathy Green own this **Stone County** barn that was built in the 1970s and is located on **Big Springs Road**. This photo was taken in October 1998 when the fall foliage was reaching its peak. *(Submitted by Kathy Green, Concord FFA Chapter)*

Descriptions of the barns pictured on the cover can be found on the following pages:

page 122

page 31

page 29

page 142

page 152

page 151

THE DONNING COMPANY PUBLISHERS
184 Business Park Drive, Suite 206
Virginia Beach, VA 23462

Steve Mull, *General Manager*
Barbara B. Buchanan, *Office Manager*
Pamela Koch, *Editor*
Andrea L. W. Eisenberger, *Graphic Designer*
Amanda Guilmain, *Imaging Artist*
Mary Ellen Wheeler, *Proofreader*
Lori Kennedy, *Project Research Coordinator*
Scott Rule, *Director of Marketing*
Travis Gallup, *Marketing Coordinator*

Ed Williams, *Project Director*

Library of Congress Cataloging-in-Publication Data

Collins, Angela B.

Barns of Arkansas / by Angela B. Collins.
p. cm.
ISBN 1-57864-286-8 (hard cover : alk. paper)

1. Barns—Arkansas—Pictorial works. I. Title.
NA8230.C65 2004
728′.922′09767—dc22
2004019471

Printed in the United States of America at Walsworth Publishing Company

Foreword

Barns: they lie on the fringes of your conscience.

Many of them are just part of your everyday landscape. You accept them and move on. When asked, you may not even remember when you saw that last barn. Why is that?

Everyone today seems to be in a hurry. We need to get to work or take the kids to their next practice. Who has time to relax, much less notice what is going on around them? When it's harvest time, we don't acknowledge the majesty of the process. We wonder, instead, about how long that combine is going to hold us up. We seem to have lost our connection with agriculture. We see that loss in many ways: our children don't know where their food comes from; we protest timber cutting yet build new homes and use paper products every day; and we can't see the basic interconnectivity of life.

We have removed ourselves from the past. We spend all of our time focusing on technology and making our lives easier when, in reality, we are complicating things. Our moms and dads, grandmas and grandpas, and others lived in a simpler time. Things were not easy, but they were less complicated. People took the time to see the majesty in harvesting. They knew intimately where their food came from. And they knew and respected the interconnectivity of all things.

They have left, for us, gentle reminders of the past and their hard work and contributions to life. We can see those reminders every day. We must choose to look.

Barns: they lie on the fringes of your conscience. But they don't have to. Look at them. Look at what they represent. Hear the stories they tell. Ask people about them. You will be amazed at what you find. ■

Acknowledgements

Barns of Arkansas is a collaborative effort between the Arkansas FFA, its members and supporters, and the Electric Cooperatives of Arkansas. When we first started discussing this publication, we were overwhelmed. Where would we find all the barns we needed? Well, we stopped and took a look around. Barns dot our landscape. They are everywhere. In fact, there were so many, we were faced with a new challenge. How were we going to capture all of these barns and find out about them in such a short time? Well, we solved that problem, too.

FFA members, their agriculture teachers, and their supporters were challenged to help us create this book. They rose to the challenge. Actually, there were a few who rose to the challenge and kept climbing! Without people like Geneva Wallace, Wynne FFA member Jessica Wallace's mom; Kathy Green, supporter of the Concord FFA chapter; and Ken Crocker, a professional barn photographer from Texas, this book would not be a reality. Special thanks to them for contributing 45 percent of the photos in the book.

We would be remiss if we did not acknowledge our support at home. My husband, Coo Collins, is a forester. He knew how important this book was to me. As he traveled the state, he would note barns and their locations. He was also very patient when, on a road trip, I would pull over to look at and photograph barns. Mr. Marion Fletcher's wife, Eva, director of Camp Couchdale, also grabbed this project by the horns and contributed many entries.

Altogether, we had 298 entries from thirty-one chapters and seven individuals. Not too bad for a first-time endeavor when everyone is so busy! What we realized is that many share our passion for barns. There are barn lovers all over Arkansas.

Barn owners told us stories about how folks line up to take pictures of their barns. Others have stories about special times in their barns including weddings and get-

togethers. Still others can tell you who built the structure and how it was passed down through the family. Even those who acquired the barn and don't know its history will tell you they "just like it." Each barn in this book is special. Each has its own story to tell. Take the time to listen to each one and enjoy!

Above: *The Jeff and Orvill Gammill families built the Varela barn in 1900. It is located in* ***Stone County*** *and is owned by Dr. Charles Varela. The barn has some unique features including the diamonds located in the loft area. This over one-hundred-year-old barn is in great shape. (Submitted by Kathy Green, Concord FFA Chapter)*

Arkansas FFA

In 1917, Congress passed the National Vocational Education Act, which established agriculture classes in high schools. Arkansas' agriculture teachers began forming "clubs" soon thereafter (Corn Club, Farm Boys). Just as Arkansas started organizing as a state, the call came from the national level that a formal organization should be chartered. Thirty-three delegates met at the Baltimore Hotel in Kansas City on November 20, 1928, to launch the Future Farmers of America. Arkansas was the second association chartered. Since its beginning, Arkansas has been involved at the national level of the FFA. The first Star Farmer, Mr. Carlton Patton, was from Arkansas; several national FFA officers have hailed from Arkansas; and the Arkansas state staff has been involved with the National Board, the Board of Trustees, and other leadership positions. The Arkansas FFA started as an organization dedicated to preparing young men for a future in farming; it has evolved into an organization dedicated to preparing all students for careers in agriculture and leadership.

The Arkansas FFA has challenged its members to Live the Legacy of FFA every day. FFA allows young people to develop on three levels: premier leadership, personal growth, and career success. We believe that our members evolve into well-rounded productive citizens.

FFA members have many opportunities to succeed. They participate in many leadership events: parliamentary procedure, opening and closing ceremonies, creed speaking, prepared speaking, and extemporaneous speaking. Other options for success include Career Development Events, which are team-oriented. Students have a tremendous selection, ranging from Livestock Evaluation to Farm Business

Management. Whatever a student's interest, there is a Career Development Event that will build on it.

Members may succeed individually as well. All students in agriculture develop and maintain a Supervised Agricultural Experience (SAE). SAEs allow students the opportunity to explore possible career pathways through entrepreneurial or placement activities. Students with excellent SAEs are recognized in the Proficiency Award process. Students can be recognized locally, statewide, or nationally.

Students also participate in leadership conferences throughout the year. State Officers conduct the Arkansas FFA Leadership Conference series in the summer, reaching over five hundred people. They also conduct District Officer Training before school begins, working with the twenty-eight district officers and district advisors to bring a seamless leadership program to all members throughout the state. Other students attend Greenhand Conferences, District Leadership Conferences, Washington Leadership Conferences, and the Rural Electric Youth Tour.

Our FFA chapters also work closely with alumni and foundation members. We regularly work with sponsors and members of the agriculture community in Arkansas. One of those special relationships is with the Electric Cooperatives of Arkansas (ECA). Our camp, Camp Couchdale, sits on land donated in 1929 by Harvey Couch, an electricity pioneer. He saw the potential in FFA and its members and chose to support it. The Electric Cooperatives continues that support by sponsoring State Farmer Degrees. ECA provides a place for FFA and FFA Foundation members to meet and is a regular contributor to our annual Foundation Drive. The Electric Cooperatives of Arkansas allows FFA members and chapters as well as agriculture programs and teachers to continue Living the Legacy! ■

The Electric Cooperatives of Arkansas

DELIVERING POWER THE COOPERATIVE WAY

On April 15, 1938, the face of rural Arkansas changed forever when First Electric Cooperative Corporation turned on power to its newly built distribution system.

In ensuing years, Arkansas' electric cooperatives, along with the national Rural Electrification Administration (REA), would build thousands of miles of new lines to bring electricity to rural Arkansans. As a result of their efforts, farmers and others living in rural areas traded in their coal-oil lamps and heavy irons heated by wood stoves for the much more convenient and efficient electric appliances. They were finally able to enjoy many of the same comforts that had been available to city-dwellers for many years.

Sixty-six years later, the state's seventeen electric distribution cooperatives are still building lines and providing power to Arkansans. Together, they serve about 440,000 homes, farms, businesses, and industries in Arkansas, Louisiana, Missouri, and Oklahoma. The cooperatives' service areas cover 60 percent of Arkansas' land area and are located in seventy-four of the state's seventy-five counties. These cooperatives also own a statewide service cooperative—Arkansas Electric Cooperatives, Inc. (AECI)—and most buy power from Arkansas Electric Cooperative Corporation (AECC), a wholesale power provider. The distribution cooperatives, along with AECI and AECC, are known collectively as the Electric Cooperatives of Arkansas.

GENERATING POWER FOR COMMUNITIES

Arkansas Electric Cooperative Corporation (AECC) began operating its own power plants in 1963 after a long and hard-fought battle for the right of cooperatives to own generation facilities. The cooperatives now operate ten generating stations that produce reliable, low-cost power for members, with independence from the investor-owned utilities.

AECC has consistently invested in expansion and upgrades to its generating capacity. These improvements have allowed AECC to serve its load obligations. They have also resulted in greater efficiency and diversity of fuel use. The investments have been made in Arkansas during a period when most other utilities have made no significant increases in capacity. The plant investments have been made gradually while rates to cooperative members have actually been reduced.

THE FUTURE LOOKS BRIGHT

Just as in 1938, the cooperatives are owned by the members they serve. Unlike investor-owned utilities, the cooperatives are not concerned with making profits to please shareholders. The cooperatives' only concern is for the members who own them.

Although the cooperatives have grown and evolved to meet the needs of their members, they continue to have faith in their member-owned system. It is a unique, grassroots system that has stood the test of time. Throughout the cooperatives' history, they have shown their commitment to members and their communities. And that will continue in the decades to come as the cooperatives help light the way for a brighter future for Arkansas. ■

Electric Cooperatives of Arkansas

Your Local Energy Partners

www.ecark.org

Your Touchstone Energy® Cooperatives

Above: Originally part of the Schug Dairy Farm, the McGrew barn can be found in **Brookland**. The barn has an old loft for storage while the bottom half of the barn is used to hold cattle. *(Submitted by Lane Lenderman, Brookland FFA Chapter)*

Right: This classic barn was built in the 1940s and was part of the Rex Dickion Plantation in **Brookland**. An interesting feature of this structure is its pulley located above the loft door. The barn currently houses cattle and hay. *(Submitted by Lane Lenderman, Brookland FFA Chapter)*

Right: This barn was built in the 1940s and was part of the Rex Dickion Plantation in **Brookland**. It features a corncrib inside with sheds on each side. The original doors were very short and narrow. Today, the barn is used to stack hay and provide shelter for horses. *(Submitted by Lane Lenderman, Brookland FFA Chapter)*

Below: The Clayton barn in **Brookland** was formerly part of the Schug Dairy Farm. This barn is used to store hay and tractors. It has been recently re-roofed and has a classic wind vane on top. *(Submitted by Lane Lenderman, Brookland FFA Chapter)*

Weathered from top to bottom, this is one of several barns on the Robert Stroud farm. The barn was built in the early 1920s, stands twenty-five feet high, and is used to store round hay bales. *(Submitted by Grant Crawford, Searcy FFA Chapter)*

The Delbert Harris barn, located near **Pangburn**, in **White County**, has fallen into a state of decay. While the old farm site is being rented out, the barn is not being used. This barn is fifty feet across, forty feet in depth, and thirty feet tall. *(Submitted by Grant Crawford, Searcy FFA Chapter)*

The Rick Marsh barn, built in January 1960, is located in **White County** near **Pangburn** and is used primarily for storage. Its dimensions: forty-five feet across, thirty feet deep, and twenty-five feet tall. *(Submitted by Grant Crawford, Searcy FFA Chapter)*

This barn is located in **Midway**, just ten miles north of Mountain Home. *(Submitted by Coty Faulkenberry, Mountain Home Jr. FFA Chapter)*

The Robert Stroud barn is located in **Desha**. This barn was built in the early 1920s and incorporated several unique concepts. The hayloft was equipped with a sliding door that utilized cables, pulleys, and counterweights to raise and lower the door as needed. Pipes with holes drilled in them were passed through the corncrib to facilitate the drying of the corn. Mr. Stroud still raises cattle today; the barn is used for storage of feed, equipment, and other items. *(Submitted by Grant Crawford, Searcy FFA Chapter)*

Above: Located in **Independence County** just outside of **Batesville**. *(Submitted by Scott Phelps, Mountain Home Jr. FFA Chapter)*

Right: Located on **West Road/ Highway 178** in **Mountain Home**. *(Submitted by Coty Faulkenberry, Mountain Home Jr. FFA Chapter)*

This **Baxter County** hay barn and matching horse barn are owned by Daffron Farms in **Mountain Home**. *(Submitted by Coty Faulkenberry, Mountain Home Jr. FFA Chapter)*

An aluminum barn can be seen at **Beffords Hill** in **Baxter County**. This Mountain Home barn, a multipurpose structure, is located on a Limousin cattle farm. *(Submitted by T. J. Frein, Mountain Home Jr. FFA Chapter)*

Mountain Home, Baxter County. *(Submitted by Kolin Paulk, Mountain Home Jr. FFA Chapter)*

This large, old, abandoned barn sits off the highway in **Batesville, Independence County**. *(Submitted by Scott Phelps, Mountain Home Jr. FFA Chapter)*

Timbo. *(Submitted by Allen Sanders, Marshall FFA Chapter)*

This old log barn is located on the Clay Diffey property in **Stone County**. The barn was built in 1945 and is owned by L. C. Gammill. It was photographed in January 1996 and is not being used today. *(Submitted by Kathy Green, Concord FFA Chapter)*

This page and opposite: Jasper Cooper built the Avey barn in the late 1930s. This red barn is still in use; Stacy Avey uses the facility for horses and hay storage. The barn is weathered and worn but will stand the test of time. *(Submitted by Kathy Green, Concord FFA Chapter)*

Located on the Jim Penn farm in western **Lawrence County**, this barn was built in the 1940s. It has been used for dairy farming, corn storage, feed grinding, cow and calf working, and hay storage. The barn is not being used today. *(Submitted by Whitney Penn, Lynn FFA Chapter)*

Troy Kendrick built this barn in 1926. It is located in **Stone County** off **Highway 66**, and Troy Morrow currently owns it. The barn is striking to look at any season of the year. Originally photographed in October 1998, it was revisited in February 2004. Mr. Morrow has a few horses that call this barn home. *(Submitted by Kathy Green, Concord FFA Chapter)*

In the **Blue Mountain** community of **Stone County**, the Ralph Roberts barn sits nestled at the base of the mountains. The Broyles family built the barn in the 1930s. Kin Roberts still uses the barn today. February is a good time for this old barn to see a little wintry weather. *(Submitted by Kathy Green, Concord FFA Chapter)*

The Dearien barn in **Stone County** has a classic style. It was built in 1942 by James Moody and is owned by his daughter, Muriel Dearien. In October 1995, the barn was flanked by fall foliage. In February 2004, the mountain looks bare with its winter layer of snow. This barn has certainly stood the test of time. *(Submitted by Kathy Green, Concord FFA Chapter)*

Built in 1955, the Sam Moore barn is located on **Pleasant Hill Road** in **Stone County**. Dee Fulks, Sam Moore's grandson, still uses the barn. Originally photographed in October 1995, by February 2004, the barn stood under snow and was beginning to show some signs of wear around the front entrance. *(Submitted by Kathy Green, Concord FFA Chapter)*

The Charles Long barn was built in the mid-1940s. In October of 2002, the barn was a bright red color. Just a year and a half later, the barn had lost most of its brightness. The Charles Long barn is located in **Stone County** on **Highway 66**. *(Submitted by Kathy Green, Concord FFA Chapter)*

Left: Built in the 1920s, this **Stone County** barn is located on **Pony Peak Road**. Oral Richardson owns it. The barn was photographed in 1997. *(Submitted by Kathy Green, Concord FFA Chapter)*

Below: Built in the late 1950s by Paul Miller, this barn is located on Tim and Liz Mullins' property in **Newnata (Stone County)**. The barn is currently used for storage. *(Submitted by Kathy Green, Concord FFA Chapter)*

Bob Miller built this barn in the 1970s near the **Blue Mountain** community in **Stone County**. Dee and Kathy Fulks own it. This photograph was taken in January 1998. *(Submitted by Kathy Green, Concord FFA Chapter)*

This barn is located in **Independence County** between **Cord** and **Charlotte**. *(Submitted by Angela B. Collins, Arkansas FFA State Executive Secretary)*

The Mountain Home FFA livestock barn was built in 1983. In 1995, an addition was added. The barn holds nineteen show steers and has wash racks, digital scales, and a feed room. Agriculture instructors are Roger Steele, Josh Baker, and Jacquie Albright. *(Submitted by Jamie Clayton, Mountain Home Jr. FFA Chapter)*

In January 1996, snow covered this magnificent barn, which is located on the Harold and Brenda Brewer farm in **Stone County**. The barn was built in the early 1900s near **North Mill Creek** and is still in use today. *(Submitted by Kathy Green, Concord FFA Chapter)*

This barn is located on **Allen Chapel Road** near **Salado**, in **Independence County**. *(Submitted by Jeremy Robertson, Advisor, Bradford FFA Chapter)*

Above: Located nine miles north of **Mountain Home**, the "Leaning Barn" was built in 1918. Noel and Elsie Crawford, owners of the "Leaning Barn," said it was used for many years as a reference when giving directions. The barn was torn down in 1998. *(Submitted by James Rhein, Mountain Home FFA Chapter)*

Right: Located on **Purcell Road** in **Paragould**, this beautiful old barn was built in 1921 by Mr. Paris Purcell. The barn measures sixty feet by one hundred feet and was built as a dairy barn. Mr. Purcell owned his own sawmill and sawed all the lumber for the barn. The materials that had to be purchased, along with some hired labor, brought the original cost to four thousand dollars. Mr. Purcell used the barn for dairy production until the 1960s. He remodeled in 1975 with new siding and windows and currently uses the barn for storage. The hayloft still has a working hay rail. *(Submitted by James Potter, Advisor, Paragould FFA Chapter)*

Above: The Patterson Farm is located in **White County**, near **Rosebud**. This classic barn was built in 1925 and is the only barn on the Patterson Farm. The Pattersons still use the barn in their cattle operation. *(Submitted by Hayden Henson, Quitman FFA Chapter)*

Left: This old wagon shed, located on the Charlie Farris homestead in **Stone County**, was photographed in October 1998. *(Submitted by Kathy Green, Concord FFA Chapter)*

Old meets new just east of **Antioch**. A new metal barn flanks an older wood barn in **White County**. Their owners, who raise Tennessee walking horses and Labrador retrievers, use both barns. *(Submitted by Nina Crutchfield, Advisor, Searcy FFA Chapter)*

This new barn is located in **White County** just west of **Beebe**. The facility is metal and was designed to hold small horses or ponies. There is also an area for small equipment storage. *(Submitted by Nina Crutchfield, Advisor, Searcy FFA Chapter)*

The Crutchfield barn was built in the summer of 2002. This thirty-six-by-thirty-six-foot pole barn has two horse stalls, a hay storage area, an indoor dog kennel, and an indoor wash area. The barn sits on twenty acres of lush horse pasture and is fronted by a pond. *(Submitted by Nina Crutchfield, Advisor, Searcy FFA Chapter)*

This weathered red and white beauty sits on top of a ridge overlooking **El Paso Valley** on one side and an orchard on the other side. The owners breed and show horses. *(Submitted by Nina Crutchfield, Advisor, Searcy FFA Chapter)*

This barn, located in **White County**, east of **Floyd**, was built in the late 1990s. A horse barn, it houses quarter horses. *(Submitted by Nina Crutchfield, Advisor, Searcy FFA Chapter)*

Located on the corner of **Black Jack Mountain Road** and **Wayne Walker Road** in **White County**, this great old barn sits proudly. The barn was built in the 1950s and is not used much today. *(Submitted by Nina Crutchfield, Advisor, Searcy FFA Chapter)*

Worn and faded, this barn houses an antique car. The barn is located just off **Highway 5** in **White County**; its property is posted as a Commercial Game Bird Refuge. *(Submitted by Nina Crutchfield, Advisor, Searcy FFA Chapter)*

Between **Center Hill** and **Floyd**, in **White County**, you can find this majestic old barn. It was built over seventy years ago and is used for storage. *(Submitted by Nina Crutchfield, Advisor, Searcy FFA Chapter)*

Located on the Jim Penn farm in western **Lawrence County**, this barn was built in the 1980s. It is used to store hay, work cows and calves, and wean calves. *(Submitted by Whitney Penn, Lynn FFA Chapter)*

Located on the Jim Penn farm in western **Lawrence County**, this barn was built in the 1930s. It has been used for grinding, storing, and feeding corn. It is still used to work cattle and store hay. *(Submitted by Whitney Penn, Lynn FFA Chapter)*

Located on the Jim Penn farm in western **Lawrence County**, this barn was built in the 1950s. It has been used for working cows and calves, weaning calves, raising hogs, and storing hay and feed. *(Submitted by Whitney Penn, Lynn FFA Chapter)*

Above: This "thoroughly modern" barn was built in 2003 and is owned by Mark Herrington. The barn is located west of **Sidney** in **Izard County**. It has a loft on the top and gateways on each side. A sharp white fence surrounds the driveway and barn. *(Submitted by Shana Herrington, Mt. Pleasant FFA Chapter)*

Right: Trees and an old wooden fence surround this old barn with its rusted roof. It rests in the middle of a field about five miles west of **Sidney** in **Izard County**. The barn has been there for many years. *(Submitted by Shana Herrington, Mt. Pleasant FFA Chapter)*

Jonesboro and **Craighead County** are in a stage of growth. Everywhere one turns, the old is coming down and the new is rising up. This classic barn, which stood on **Caraway Road**, no longer graces the landscape. It has been razed and replaced with a Western Sizzlin. *(Submitted by Joyce Sandy)*

Above: On **Highway 351**, just north of **Jonesboro**, the spring daffodil-watchers visited this barn yearly. This graceful sentinel no longer stands vigil. *(Submitted by Joyce Sandy)*

This page and bottom of opposite page: Battered and beaten, this barn has survived for sixty-seven years. It was built in 1937 north of **Augusta** next to the **White River**. The barn has changed hands several times. The mound in front of the barn (right) was used as a livestock ramp. An old piece of equipment has been stored there (below) for years. *(Submitted by Brandy Fortune, Augusta FFA Chapter)*

Left: The Garrett barn is located within the city limits of **Augusta** in **Woodruff County**. The Garretts sold the barn to Jim Moore, who now raises chickens there. *(Submitted by Kendrick Briscoe, Augusta FFA Chapter)*

Below: Jerry and Patsy Cook own a little place on **Toombs Hill** in **Clay County**. This photo, taken in the spring of 1991, shows a fabulous little barn. The barn was built about seventy-five years ago and is used to store hay and feed on their livestock facility. *(Submitted by John Golden, Augusta FFA Chapter)*

Above: This whitewashed barn was built in 1948 in **Providence**. The timber on the property was cut and sawed on-site. *(Submitted by Randy Kidd, Augusta FFA Chapter)*

Left: William "Bill" Kinsey owns this barn, which is located in the **White River** bottoms one-quarter of a mile from an Electric Cooperative plant. The barn is over fifty years old and has seen its share of storms. *(Submitted by Matthew Hunter, Augusta FFA Chapter)*

The old **Arkansas State University** dairy barn as it appears in the 1939 ASU yearbook. The barn stood where the post office now stands. *(Submitted by David Agnew, Arkansas State University)*

Max Dacus owns this barn and has exerted great effort to maintain it. The barn has weathered several storms and has been repaired after each. A pre–Civil War log house stood on this site for many years. *(Submitted by David Agnew, Arkansas State University)*

Lester Dacus built this **Craighead County** barn in 1939–1940. *(Submitted by David Agnew, Arkansas State University)*

This barn dates back three generations. Mrs. Hathcoat, who lives on the old homeplace, recalls that her grandfather, whose last name was Burroughs, was the first owner of the property. The barn was damaged a few years ago by a storm. *(Submitted by David Agnew, Arkansas State University)*

Craighead County, north of **Jonesboro**. *(Submitted by David Agnew, Arkansas State University)*

This barn, located between **Brookland** and **Jonesboro**, is over eighty years old and was the homeplace for Olen Ackers. This is also on the homestead where Lonnie McNatt, former Arkansas Director of Vocational Education, grew up as a boy. *(Submitted by David Agnew, Arkansas State University)*

This old log barn is on a piece of property just north of **Jonesboro** owned by Larry and Ruby Hinson. They have only lived there since 1994 but know from land records that the barn was there before 1930. The barn is held together with wooden pegs, and the wagon on the south side, made by John Deere, still has its original paint. *(Submitted by David Agnew, Arkansas State University)*

At a cost of $100,000, the **Arkansas State University** dairy barn was built in 1943. The barn stood on the east side of campus near the area where the Convocation Center is located. Partly destroyed by fire in 1986, the barn was removed in 1988. *(Submitted by David Agnew, Arkansas State University)*

Lacey Bilbrey built this barn around 1925; his descendents, Randy and Karna McFadden, now own it. The barn sits atop a hill and is a landmark in the **Imboden** area. *(Submitted by Terry Rorex, Advisor, Black Rock FFA Chapter)*

In 1904, a renowned **Lawrence County** official built this barn. After it was built, he gave the barn to his daughter as a wedding gift. The Meadows family restored the barn in 1976. *(Submitted by Chris Propst, Black Rock FFA Chapter)*

The Tate-Propst barn was built in 1946 using materials from a much larger barn that had stood in its place some years before. Lester Hankins built the barn with help from his son. The barn has now been in the Tate-Propst family for over half a century. *(Submitted by Chris Propst, Black Rock FFA Chapter)*

Bo Pierce of **Ravenden** built the Pierce barn in 1949. It was built of cypress and oak with dimensions of seventy feet by seventy feet with a ten-foot stone wall. The current owner is Cloyece Burrow. *(Submitted by Chris Propst, Black Rock FFA Chapter)*

The Shannon barn is located between **Black Rock** and **Powhatan**, in **Lawrence County**. Louie Pruitt built the barn in 1930 for J. E. Krone. On a twenty-one-inch two-by-four, Pruitt noted materials that would be needed; he kept that piece of lumber for years. Eldon Shannon, who still owns the barn, restored it in the 1970s. *(Submitted by Chris Propst, Black Rock FFA Chapter)*

Originally built to house livestock and store feed and hay, this **Independence County** barn now sits vacant. *(Submitted by Cushman FFA Chapter)*

This old barn is located on **Medlock Road** in **Independence County**. It has a loft and is used to store hay. *(Submitted by Cushman FFA Chapter)*

Above: This barn, located in **Independence County**, is used to store round bales of hay. It was originally built to store square bales and to feed and house livestock. *(Submitted by Cushman FFA Chapter)*

Right: Payton Daniels built this barn in 1960 in **McClelland (Woodruff County)**. The facility was used to raise cattle and is still in use today. *(Submitted by Jimmy Skinner, Augusta FFA Chapter)*

Above: Located just south of **Cushman**, this **Independence County** barn is used to provide shelter for show steers and to store hay. *(Submitted by Jeffrey Baxter, Cushman FFA Chapter)*

Left: This **Independence County** barn was originally built to store hay. It is now mainly used for storage. *(Submitted by Jeffrey Baxter, Cushman FFA Chapter)*

Left: Round bales of hay are now stored in this **Independence County** barn, originally built for beef cattle. *(Submitted by Brittany Hurley, Cushman FFA Chapter)*

Below: This **Independence County** barn is for horses. It has four stalls, holds about two hundred square bales of hay, and has a tack room for accessories and essentials. *(Submitted by Brittany Hurley, Cushman FFA Chapter)*

The workmanship and style of this barn is timeless. If you are driving on **Highway 9** from **Mountain View** to **Shirley** (west), this classic wooden hay barn can be seen in a hollow. Coming in from Shirley, you can't see it. These photos were taken in early October 1998 and showcase the barn's elegance amidst the fall foliage. The barn is estimated to be eighty years old. *(Submitted by Ken Crocker)*

Located between **Heber Springs** and **Searcy**, this barn has character. The tin roof was put on prior to the spring of 1997 to keep the elements out and the hay dry. *(Submitted by Ken Crocker)*

My Dad, brother, and I built this barn in 1999. It has stalls, a tack room, and a feed room on the west side. The east side has portable panels to make pens for horses, cattle, show lambs, equipment, and hay storage. The barn has a forty-foot-by-sixty-foot free span in the center for riding. Outside the barn are a drag steer (roping dummy), a lamb trim table, and a rodent control system (the cat). The silhouette in the center is my brother, BJ. *(Submitted by Lane Lenderman, Brookland FFA Chapter)*

This page: Any season of the year, this classic barn in **Jonesboro** is a sight to behold. Located in **Craighead County** on a rise, the barn is used in a cattle operation. *(Submitted by Amber Keefer, Nettleton FFA Chapter)*

Above: Located in **Independence County**, this barn was built around 1948 by Lacy Wilf. Wilf operated a steam-engine sawmill. He and his workers sawed the lumber to build the barn. He used this barn for his horses and to store hay. Mr. Gale Thomas now owns the barn. *(Submitted by Geneva Wallace, Wynne FFA Chapter)*

Right: Mert Nevell owns this barn, built in 1948 for Millard Burris. It was built for the family's team of mules and a saddle mare. There were three stables (each animal knew its own), a corncrib, and a hayloft. Many in the area believe Lacy Wilt helped construct this barn. *(Submitted by Geneva Wallace, Wynne FFA Chapter)*

Time has taken its toll on this **Independence County** barn. Mr. and Mrs. Luther Satter of Pleasant Plains built the barn in 1934. Its roof is fading, and its boards are showing wear. Nonetheless, this barn is majestic. *(Submitted by Geneva Wallace, Wynne FFA Chapter)*

Right: Fred Ferguson built this barn in 1950 to raise cattle and store hay. His son, Kenneth, now owns the barn. Kenneth rents it out to Chris Tharp, who uses the barn to grind feed. *(Submitted by Geneva Wallace, Wynne FFA Chapter)*

Below: In 1950, James Harold Wilf (better known to his friends as "Mutt") built this barn, located in **Pleasant Plains**, to store his tools. Mrs. Wilf still owns the barn today. *(Submitted by Geneva Wallace, Wynne FFA Chapter)*

Above: Johnny Bryant built this interesting barn in 1939. He cut his own lumber, vertically sawing his pine rafters. This type of cut produces very durable lumber. Natural stone is used as pilings to support the structure. *(Submitted by Geneva Wallace, Wynne FFA Chapter)*

Right: Kenneth Swaim built this barn around 1948 with the help of his dad (Saul) and his brother (Gerald). They cut the logs and sawed the lumber themselves. The roof was originally shingled with wooden shingles and later replaced with sheet iron. The barn was used to store hay. Kenneth's daughter and son-in-law (the Galloways) now own the barn. *(Submitted by Geneva Wallace, Wynne FFA Chapter)*

Left: Saul Swaim built this **Independence County** barn in 1952 with the help of both his sons, Kenneth and Gerald. They cut the logs and sawed the lumber themselves. *(Submitted by Geneva Wallace, Wynne FFA Chapter)*

Below: In 1950, Mr. Ardis Hughes cut down some trees on his land, had them sawed into lumber, and built this barn. He used the barn for his cattle and for hay storage. It is now owned by his niece, Dixie Gunner, and is used for her horses. *(Submitted by Geneva Wallace, Wynne FFA Chapter)*

This interesting barn was built in the early 1930s on the homeplace of Jerry Dobbs. It has a huge loft and several stalls. The barn looks to be in disuse; however, it is wired for electricity and kept clean. The roof is also in good shape. *(Submitted by Geneva Wallace, Wynne FFA Chapter)*

Amos Scantlin had this **White County** barn built in 1950. It is unique in that the outside of the barn does not look like typical barns in the area. Jim and Natalie Dunham own the barn now and use it for their horses. *(Submitted by Geneva Wallace, Wynne FFA Chapter)*

William "Bill" Paul Ritzel built this barn in the late 1960s to be a center point for a beef herd. It was built on the 1886 homestead of Harvey, Heart, and Ritzel. Bill's parents, Hank and Wilma, raised him on the old homeplace. Weather and time have taken their toll on this once majestic structure. A couple of tractors and harvesters are still stored here. *(Submitted by Geneva Wallace, Wynne FFA Chapter)*

Billy and Lois Ward built this barn in 1956. They used the barn for their cows, geese, and hogs. Susan Reed now owns the barn. *(Submitted by Geneva Wallace, Wynne FFA Chapter)*

Mr. and Mrs. John Rhew own this **White County** barn. In 1932, when Mrs. Rhew was ten years old, her father built the barn. The family used the barn to feed three or four cows and a couple of horses and to store the feed and hay they had grown. *(Submitted by Geneva Wallace, Wynne FFA Chapter)*

Clarence and Nell Etheridge built this **White County** barn in the early 1940s. They used it for their cattle and for hay storage. Their daughter, Willie Cantrell, now owns the barn. *(Submitted by Geneva Wallace, Wynne FFA Chapter)*

The Clifton March barn was built during the Depression. It was a cattle barn that all the neighborhood children used for play. They would swing from the rafters and play in the loft. Over half of the tin on one side of the roof is missing now, but this old barn still holds memories for many people. *(Submitted by Geneva Wallace, Wynne FFA Chapter)*

Alvin Gordon built this **White County** barn in 1945. He used it to raise cattle, chicken, and other farm animals and to store hay. His son, Jimmy, a retired agriculture teacher, now owns it. The Gordon brothers use the barn to stable horses and to store farm machinery. *(Submitted by Geneva Wallace, Wynne FFA Chapter)*

This barn was built by Ray and Velva Cross around 1957. The Crosses used it for their cattle and to store hay. They shipped butter (made from their cow's cream) to St. Louis to sell, five gallons at a time. Their son, DeWayne Cross, now owns the barn. *(Submitted by Geneva Wallace, Wynne FFA Chapter)*

Thurman and Vera Turner built this barn in the early 1960s. They used it for cattle and hay/feed storage. From the back of the barn, you can see the hay chute. Winifred Simpson now owns this classic. *(Submitted by Geneva Wallace, Wynne FFA Chapter)*

Golden and Helen McCartney built this red barn around 1950. They used it primarily for their cattle and for hay storage. A couple of pieces of new tin grace its stately rust red roof. While it is a little overgrown, this barn maintains its dignity. *(Submitted by Geneva Wallace, Wynne FFA Chapter)*

Above: This majestic wooden barn was built in 1944 on the Floyd Moore homeplace. Moore owned his own sawmill and cut his own lumber. He raised cattle and horses and then converted the barn for sheep production. His daughter, Helen Simpson, now owns the barn. She recently built a new metal barn in which to raise her sheep. *(Submitted by Geneva Wallace, Wynne FFA Chapter)*

Right: When L. G. Kendall retired in 1974, he decided to raise cattle and build a barn. This barn has a concrete foundation, a tin roof, and cypress lumber siding. L. G.'s wife, Veo, now owns this **White County** barn. *(Submitted by Geneva Wallace, Wynne FFA Chapter)*

Walter Grundon's barn was built in the 1930s. He used it to raise cattle, pigs, and other farm animals. The Grundon farm was well known in the area. People would come from all around to swim in the big pool with the screened-in porch and bathhouse and to dance on the dance floor. In 1960, A. B. "Blue" Clark bought the farm and added a watermill (left) to generate electricity. John White Jr. now owns the farm. *(Submitted by Geneva Wallace, Wynne FFA Chapter)*

This **Poinsett County** barn was built on the Clay Mardis farm in the early 1940s. Mardis used it to store hay and feed and to shelter cattle and horses. The Peck family bought the farm later and left the barn standing. *(Submitted by Geneva Wallace, Wynne FFA Chapter)*

Fred Moulder built this barn in 1948 in the **Bay Village** community of **Poinsett County**. He stored hay in the loft, fed his cattle inside, and stored corn in the corncrib. His son, William, now owns the barn with his wife, Lois. *(Submitted by Geneva Wallace, Wynne FFA Chapter)*

In 1942, the Dean Massengill family built this big white barn in **Poinsett County**. They used the facility for cattle and horses. The barn is used today to store hay and to work cattle. It is in very good shape. *(Submitted by Geneva Wallace, Wynne FFA Chapter)*

Vernon Mason and Ray Alexander rebuilt the old Crisco place at **Sunnyside** in 1952. They used lumber from the first barn on the place. That lumber was over one hundred years old. Earl Ransom owned the barn and has passed it down to his daughter and her husband, the Scoggins family. *(Submitted by Geneva Wallace, Wynne FFA Chapter)*

This **Woodruff County** barn is located south of **Morton**. Loyce Mitchell built it in the 1970s out of scrap cypress lumber from an old house. Mitchell used the barn to store hay. He also fed his cattle from the feed troughs located in the wings of the barn. *(Submitted by Geneva Wallace, Wynne FFA Chapter)*

Around 1904, E. E. Heffington built this barn in **Heffington** (now **Bradford**). His son, W. H. Heffington, inherited the barn after his father drowned in 1907. The barn then passed to B. W. and Nannie Heffington in 1929. In 1959, it became the property of Cyrus Allen and Della Warren. They used it to raise sheep. After passing through the Farmer family, the barn currently resides with Kenneth and Norma Sexton, who use the facility to raise cattle and to store hay. *(Submitted by Geneva Wallace, Wynne FFA Chapter)*

Ernest "Bud" and Anna Faye McAdams own this barn in **Jackson County**. It was built between 1939 and 1941 by Anna Faye's parents, Herman and Vera Lody. W. C. Ward sawed the lumber at a sawmill, and Henry and Howard Solida built the barn. George Pearce and N. C. "Red" McAdams hauled the logs. The lumber is cypress. The Lody family used the barn for their cattle and to store hay and feed. There was a place in the barn to milk cows. The loft still has its original hayfork, and the corncrib is intact. *(Submitted by Geneva Wallace, Wynne FFA Chapter)*

Above: Ivan and Nell Anderson own this **Stone County** barn. Ivan's father, Floyd Anderson, and his grandfather, Will Anderson, built this barn around 1935. They cut the logs themselves, sawed the lumber at a sawmill, and bought the nails and roofing. The barn was used to store loose hay in the spring; around July, they would push it out and bale it with a hand-fed horse-drawn baler. The barn also provided shelter for their four mules. Bats and barn swallows now occupy the barn in the summers. *(Submitted by Geneva Wallace, Wynne FFA Chapter)*

Left: Between the towns of **Augusta** and **Gregory**, Nathan Gregory owns a cattle farm and, on it, this lovely barn. Nathan has owned the farm since 1978; it was passed to him from his father. Prior owners include Nathan's grandmother (Mary Noel Gregory), his grandfather (Minor), and his great-grandfather. This is truly a family farm. *(Submitted by Tonya Nichols, Augusta FFA Chapter)*

Agness Ball owns this classic red barn in **Woodruff County**, just off **Highway 64**. Her father, Joe Hess, built this barn in the 1950s. The Hess and Ball families are well known in the area. The barn is used today in conjunction with a rice-farming operation. *(Submitted by Geneva Wallace, Wynne FFA Chapter)*

This formerly red barn in **Woodruff County**, just three miles north of **Gregory**, was built in the 1940s, after the war. Paul Collins now owns the Gregory Farm, which was once used for cattle and horses. *(Submitted by Geneva Wallace, Wynne FFA Chapter)*

Buel Gilbert built this **Stone County** barn around 1945 for his cattle and horses. He also stored hay in it. His daughter and her husband, Debra and Foy Eddings, now own the barn. *(Submitted by Geneva Wallace, Wynne FFA Chapter)*

The Burks family, in **Cross County**, started a Tennessee walking horse business and needed a barn. Responding to the need, Mr. King built this barn in 1973. The Burks raised world grand champions here. Mother Nature is slowly reclaiming this barn, and it is blending in with the landscape. *(Submitted by Geneva Wallace, Wynne FFA Chapter)*

Dr. Willard Burks had this **Cross County** barn built by Howard Walker in 1976 when the family expanded its Tennessee walking horse business. Other horses also board in the barn. *(Submitted by Geneva Wallace, Wynne FFA Chapter)*

Left: Some barns are owned by the same family for generations. Others, like this red **Cross County** barn, are passed from owner to owner. Mallie Stutts Burford built this barn in 1940. Other owners included Jake Wiltshire, Bill Ransom, Gerald Stutts (in 1963), and Bobby Ellis. The barn is sitting next to prime farmland. Regardless of its ownership, it has stood the test of time. *(Submitted by Geneva Wallace, Wynne FFA Chapter)*

Below: Carl Norviel built this fabulous barn in 1950 to raise sheep. Mr. Norviel ran into a few problems, however. The area was not suitable for sheep: it was too wet and too hot, and gnats tormented the sheep. The barn has a concrete foundation and electricity and is in remarkable shape. Mrs. Faye Norviel, Carl's wife, still owns this **Cross County** barn. *(Submitted by Geneva Wallace, Wynne FFA Chapter)*

Willie E. Wilson built this **Cross County** barn around 1950. Mr. Wilson hired one carpenter and used his farm workers—unskilled in carpentry—to complete the structure. The barn was made out of cypress lumber. Although they did have electricity to run the cement mixer, all the ingredients were shoveled by hand and moved with a wheelbarrow. At one time, this barn housed around one hundred head of cattle. Mr. Wilson's son and daughter-in-law, Leonard and Joy Wilson, now own the barn. *(Submitted by Geneva Wallace, Wynne FFA Chapter)*

Above: Highway 64 is a major artery through **Cross County** and the site of prime farmland. Jess Cox built this barn in the early 1940s with poles and siding that came from the trees on his farm. Most barns are used to store equipment for crops, and this **Fair Oaks** barn is no exception. Cox used it to store his equipment and horses. The barn is still being used today. *(Submitted by Geneva Wallace, Wynne FFA Chapter)*

Left: In 1919, Fred Rolfe, a registered Hereford breeder, built this **Cross County** barn in **Birdeye** to raise his cattle. Mr. and Mrs. John H. Johnston Jr. bought the farm in 1927 and still use the facility to raise cattle. *(Submitted by Geneva Wallace, Wynne FFA Chapter)*

Right: This barn was built in the 1940s on the Morris Smith Plantation. Smith had between one hundred and two hundred sharecroppers. The plantation had its own store and post office and served as a town for the sharecroppers. Mr. Smith raised registered black angus and prize bulls worth up to an estimated $72,000 each. *(Submitted by Geneva Wallace, Wynne FFA Chapter)*

Below: Earl and Vera Peeler built this barn in the early 1940s. It was used for work animals—two mules and two horses—and milk cows. The harnesses were kept on one side. A corncrib held the white corn that was ground for meal. A pulley and hooks lifted the hay into the loft. *(Submitted by Geneva Wallace, Wynne FFA Chapter)*

Above: Shirley Willis owns this **Cross County** barn off **Highway 64**. It was built in 1948, has a classic peaked roof, and was originally used for cattle and for hay storage. The barn was later converted to raise horses. *(Submitted by Geneva Wallace, Wynne FFA Chapter)*

Right: Mr. and Mrs. Edward D. McKnight built this **Cross County** barn, located in **Parkin**, around 1945. In the years following World War II, McKnight used the barn for his mules. Later, he used the barn for his horses and cattle. Mrs. McKnight still owns the barn today. *(Submitted by Geneva Wallace, Wynne FFA Chapter)*

This old barn on **Meadow Creek** in **Stone County** is at least fifty years old. It was once used for loose hay that was lifted by hayforks on a track. *(Submitted by Jerry Shannon, Rural Special FFA Chapter)*

Left: Located two miles south of **Cave City**. *(Submitted by Geneva Wallace, Wynne FFA Chapter)*

Below: Located in **Mountain View**. *(Submitted by Clyde Dawes, Advisor, Mountain View FFA Chapter)*

Above: Located north of **Evening Shade**. *(Submitted by Geneva Wallace, Wynne FFA Chapter)*

Left: Roy Jolly of **Vandale** built this barn on the Gardner homeplace around 1928; it was enlarged to its present size in 1930. The barn was used to raise and feed Hereford cattle, which were sent to the St. Louis and Memphis stockyards. Jolly supplied meat to Bay's Meat Market in Wynne. Since the 1950s, the Leslie Nix family has farmed the property and used the barn. *(Submitted by Geneva Wallace, Wynne FFA Chapter)*

Located north of **Wynne**. *(Submitted by Geneva Wallace, Wynne FFA Chapter)*

Oscar Olson built this **Cross County** barn around 1953. His daughter, Margaret Hansen, now owns it. *(Submitted by Geneva Wallace, Wynne FFA Chapter)*

Located in **Cross County** near **Birdeye Road**. *(Submitted by Geneva Wallace, Wynne FFA Chapter)*

Located in **Cross County** off **Highway 64**. *(Submitted by Geneva Wallace, Wynne FFA Chapter)*

Located in **Bay Village** on **Birdeye Road**. *(Submitted by Geneva Wallace, Wynne FFA Chapter)*

Above: Jack Edwards built this **Cross County** barn in 1946. *(Submitted by Geneva Wallace, Wynne FFA Chapter)*

Right: This barn, built by Lewis Chappell in 1947, is located ten miles west of **Wynne**. *(Submitted by Geneva Wallace, Wynne FFA Chapter)*

William and Allene Hearyman built this **White County** barn in 1948. Jimmy Mason, who uses it to store hay and house cattle, now owns it. *(Submitted by Geneva Wallace, Wynne FFA Chapter)*

Left: Ed Davis owns this **White County** barn, built by William "Willie" Alcy and located near **Letona**. *(Submitted by Geneva Wallace, Wynne FFA Chapter)*

Below: Bob McGilvery built this barn, located just north of **Russell**, in 1949 for Clovis Crawford. Gary Little now owns the barn. *(Submitted by Geneva Wallace, Wynne FFA Chapter)*

Opposite page and right: Grover Walter built this barn in **Velvet Ridge** in the late 1930s or early 1940s. Kirk King currently owns it. *(Submitted by Geneva Wallace, Wynne FFA Chapter)*

Below: This **White County** barn has been remodeled and now serves as a residence. *(Submitted by Geneva Wallace, Wynne FFA Chapter)*

Above: Phyllis Joslin Gribsby owns the Joslin family barn, located just north of **Bradford**. She estimates the barn to be at least one hundred years old. *(Submitted by Geneva Wallace, Wynne FFA Chapter)*

Left: Located south of **Morton** in **Woodruff County**. *(Submitted by Geneva Wallace, Wynne FFA Chapter)*

Right: Located just outside of **Batesville** in **Southside**. *(Submitted by Geneva Wallace, Wynne FFA Chapter)*

Below: Ruby Roetzel owns this **White County** barn used for hay and cattle. *(Submitted by Geneva Wallace, Wynne FFA Chapter)*

Located north of **Bald Knob**. *(Submitted by Geneva Wallace, Wynne FFA Chapter)*

Above: Located just outside of **Batesville** in **Southside**. *(Submitted by Geneva Wallace, Wynne FFA Chapter)*

Left: This barn is located in **Pleasant Plains** behind the old Ernest Hook house. *(Submitted by Geneva Wallace, Wynne FFA Chapter)*

Right: Located in **Cave City**, north of the "Big Red Barn," a longtime landmark. *(Submitted by Geneva Wallace, Wynne FFA Chapter)*

Below: Located in **Mountain View**. *(Submitted by Clyde Dawes, Advisor, Mountain View FFA Chapter)*

Located in **Mountain View**. *(Submitted by Clyde Dawes, Advisor, Mountain View FFA Chapter)*

Right: This old barn is located in **Southside**, just south of **Batesville**. *(Submitted by Amanda Free, Bradford FFA Chapter)*

Below: This old barn is located in **Newark**. *(Submitted by Amanda Free, Bradford FFA Chapter)*

Above: This old barn is located in **Newark**. *(Submitted by Amanda Free, Bradford FFA Chapter)*

Right: Robert Sanford built this **White County** barn in the early 1900s. *(Submitted by Amanda Free, Bradford FFA Chapter)*

This hay barn in **Hardy** is fifty years old. *(Submitted by John Olsen Wolverton, Highland FFA Chapter)*

Above: This barn is located in **Pope County** between **Russellville** and **Dover**. *(Submitted by Patti Priest, Advisor, Ola FFA Chapter)*

Right: This barn is located on **Highway 7 North** between **Russellville** and **Dover**. *(Submitted by Patti Priest, Advisor, Ola FFA Chapter)*

Left: The "Corn Crib" was built in the 1940s. Larry and Pauline Stanberry own this barn found on **Mt. George** in **Yell County**. It was originally used as a corncrib and milking shed. The barn has post oak seals and seasoned gum siding. *(Submitted by Patti Priest, Advisor, Ola FFA Chapter)*

Below: This barn is located in **Belleville**. *(Submitted by Patti Priest, Advisor, Ola FFA Chapter)*

Left: This barn is located in downtown **Belleville**. *(Submitted by Patti Priest, Advisor, Ola FFA Chapter)*

Below: C. E. Suggs, a retired row crop farmer, owns this classic red barn. Suggs raises cattle in **Belleville**. *(Submitted by Patti Priest, Advisor, Ola FFA Chapter)*

Above: This abandoned barn in **Yell County** was originally an old house. *(Submitted by Patti Priest, Advisor, Ola FFA Chapter)*

Left: This barn is located in **Danville**. *(Submitted by Nancy Dewitt, Ola FFA Chapter)*

Carol Dewitt owns this modern-day barn, located in **Yell County** between **Belleville** and **Danville**. Carol and her family enjoy spending time there, relaxing and socializing. The Dewitts house their show horses in the facility. *(Submitted by Nancy Dewitt, Ola FFA Chapter)*

The Don Mannor barn was built in 1926 by a German immigrant and, with the exception of a few replacement sheets of metal, it still retains its original roof. The Mannors raise cattle, and the barn is used to store feed, farm equipment, and other items. This barn, located in **Cleburne County** near **Drasco**, is a popular one, with people often stopping to take pictures. *(Submitted by Grant Crawford, Searcy FFA Chapter)*

Located in **Morningstar**. *(Submitted by Troy Dye, Marshall FFA Chapter)*

Located in **Harriet**. *(Submitted by Troy Dye, Marshall FFA Chapter)*

The "Spirit Barn" is located in **Flippin**, across from Ranger Boats. Each senior class, since 1980, has added a bit of school spirit to this barn. *(Submitted by Daniel Chentnik, Mountain Home Jr. FFA Chapter)*

Above: The Ozark Mountains provide a backdrop for this scene of solitude. Buford "Red" Tatum owns the hay barn behind the pond, located in **Danville**. Mr. Tatum is a retired fisheries and wildlife professor from Arkansas Tech University. *(Submitted by Nancy Dewitt, Ola FFA Chapter)*

Above: This barn was built about 150 years ago. It was used for livestock housing and hay and feed storage on the Sinclair farm between **Pottsville** and **Atkins**. In the early 1990s, the barn was moved to Potts Inn. Potts Inn was built in the 1850s with slave labor; it was a stop on the Butterfield Stage Line and now serves as a museum. *(Submitted by Blake Hendricks, Marshall FFA Chapter)*

Right: Slowly falling apart after being abandoned, this barn was originally built to house mules. Before mechanical equipment was used on the farm, mules were the primary source of labor. This barn can be found on the Charles Oats farm just south of **Pottsville**. *(Submitted by Blake Hendricks, Marshall FFA Chapter)*

Left: This barn is located on the Joe Maxwell farm. It was built in the 1800s out of white oak logs that were hewn with an axe. *(Submitted by Blake Hendricks, Marshall FFA Chapter)*

Below: Located in **Marshall**. *(Submitted by Jonathan Bryan, Marshall FFA Chapter)*

This **Marshall** barn was built in 1928, the same year that the National FFA was formed. *(Submitted by Eden Passmore, Marshall FFA Chapter)*

Right: Located in **Marshall**. *(Submitted by Jonathan Bryan, Marshall FFA Chapter)*

Below: The Grady Horton barn was built in the late 1940s and can be found in **Snowball**. *(Submitted by Eden Passmore, Marshall FFA Chapter)*

Above: Horton's Old Barn was built in the mid-1940s and can be found in **Marshall**. *(Submitted by Eden Passmore, Marshall FFA Chapter)*

Left: The Taylor barn was built in the mid-1950s and can be found in **Snowball**. *(Submitted by Eden Passmore, Marshall FFA Chapter)*

Right: The Coulson barn is located in **Snowball**. *(Submitted by Eden Passmore, Marshall FFA Chapter)*

Below: This **Searcy County** barn was built in the late 1970s for storing hay and housing livestock. *(Submitted by Andy Griffith, Marshall FFA Chapter)*

Above: Located on **County Road 58** near **Snowball**, in **Searcy County**. *(Submitted by Bryan Goodrich, Marshall FFA Chapter)*

Left: The Firestone family built this barn, located in **Harriet**, for potato storage. *(Submitted by Kyle Yarbrough, Marshall FFA Chapter)*

Right: Located in **Harriet, Searcy County**. *(Submitted by Garrett Yarbrough, Marshall FFA Chapter)*

Below: This barn was built in the 1930s and was used for hay storage and livestock housing. Present owner Jerry Smith re-roofed and repaired it in 2002. *(Submitted by Jodie Pack, Marshall FFA Chapter)*

Jennings Dairy Barn, located in **Searcy County**, was built in 1959. *(Submitted by John Estes, Marshall FFA Chapter)*

Right: The Stephenson Dairy Barn, located near **Marshall**, was built in the 1940s. *(Submitted by John Estes, Marshall FFA Chapter)*

Below: The Tilley family owns this barn, photographed in November 1997. The structure is located in the **Landis** community of **Searcy County** and is used to store hay. *(Submitted by Kathy Green, Concord FFA Chapter)*

Highway 65, Searcy County, near **Marshall**. Taken in December 2002. *(Submitted by Kathy Green, Concord FFA Chapter)*

This **Cleburne County** barn was built in the early 1940s, has survived many tornados, and is now owned by Joyce Cannon. *(Submitted by Kathy Green, Concord FFA Chapter)*

Dave and Sandra Langston raise miniature horses on the Lazy L Ranch in **Benton County**. The interior of this gable roof–style barn has been customized to fit their needs. *(Submitted by Jenessa Arnold, Bentonville FFA Chapter)*

Thurman and Dessie Hogue built this barn in **Searcy County** near **Marshall** in 1952. It was one of the first Grade A dairies in the Marshall area. *(Submitted by Cody Hogue, Marshall FFA Chapter)*

Highway 65, Searcy County, located on the old Clay Diffey property. *(Submitted by Kathy Green, Concord FFA Chapter)*

Located in **Searcy County, Pumpkin Flat**. *(Submitted by Anthony Elliot, Marshall FFA Chapter)*

Above: Benton County, Pearl Road, abandoned barn. *(Submitted by Jenessa Arnold, Bentonville FFA Chapter)*

Right: Located in **Leslie, Searcy County**. *(Submitted by John Estes, Marshall FFA Chapter)*

This barn is located in **Parks**. *(Submitted by Dayna Manning, Waldron FFA Chapter)*

Searcy County, just outside of **Pumpkin Flat**, Jim and Nellie Harris use this barn for hay storage. The Harris family built the barn in 1960. *(Submitted by Chris Harris, Marshall FFA Chapter)*

In 1968, Jim Harris and his family built this barn in **Pumpkin Flat**. The barn was built to house chickens in the front and to milk Jerseys in the back. Today, it is used primarily for storage. *(Submitted by Chris Harris, Marshall FFA Chapter)*

In **Franklin County**, this majestic barn still stands today. The barn was built in 1909 to house horses and mules and to store hay. The loft, when not full of hay, served as a basketball court and baseball field for area youngsters on rainy days. *(Submitted by Taylor Gattis, Advisor, County Line FFA Chapter)*

This barn is located in south **Scott County**. *(Submitted by Danny Lee, Waldron FFA Chapter)*

Dave and Linda Laughlin built K-9 Camp Laughlin in 1996. They lived in an efficiency apartment on the property and built the barn around it. The apartment is sided with rough cypress slats and is now used as an indoor area for boarding dogs. Two stalls on each end, measuring twenty-four by twenty-four feet, flank that area. Half of the area is used for stall space (with additional stall space provided by pipe corrals) while the other half is used for storage of hay and equipment. *(Submitted by Nina Crutchfield, Advisor, Searcy FFA Chapter)*

If you are traveling **I-540** to **Fayetteville**, keep your eyes peeled for this beauty. If you pass it during the Christmas holidays, it will be lit up for your enjoyment. You may also be invited to one of its many wedding receptions or parties. *(Submitted by Cassie Bacon, Prairie Grove FFA Chapter)*

Above: This barn is located in **Cecler Creek**. *(Submitted by Dayna Manning, Waldron FFA Chapter)*

Left: Midway between **Paris** and **Dardanelle**, Sonny Boy's barn can be found. Oren "Sonny Boy" Horne owns the barn built in the late 1700s. Originally used as a dwelling, it had a fireplace inside. *(Submitted by Mack Wheeler, Advisor, Scranton FFA Chapter)*

Right: This tin barn, like many barns in northwest Arkansas, is used to raise beef cattle. **Benton** and **Washington Counties** boast the largest beef cattle numbers in the state. This barn is located northwest of **Prairie Grove**. *(Submitted by Cassie Bacon, Prairie Grove FFA Chapter)*

Below: **Fayetteville** is growing leaps and bounds toward **Farmington**. This classy old barn sits in an area where the fastest growing crop is new homes. Even surrounded by homes, this barn is simply majestic. *(Submitted by Cassie Bacon, Prairie Grove FFA Chapter)*

Between **Prairie Grove** and **Farmington**, this tattered old barn still stands. Its red and green combination merely hides its deeply weathered look. *(Submitted by Cassie Bacon, Prairie Grove FFA Chapter)*

This simple barn used to sit in the countryside. It now sits in the midst of several newly built homes in **Washington County**. This has become a very populated area in recent years. If you look closely, underneath the rusted tin, you can see the original wooden shingles. *(Submitted by Cassie Bacon, Prairie Grove FFA Chapter)*

Like many barns across Arkansas, this barn serves as shelter for livestock. Beef cattle use the facility to retreat from rain and snow, and the owners use the center of the barn to store hay. This barn is located in **Washington County**. *(Submitted by Cassie Bacon, Prairie Grove FFA Chapter)*

Above: Willowbark Farms is a horse production facility located in **Washington County**. The barn was originally built by an Alaskan oil driller but was recently renovated by Jerry Valentine. *(Submitted by Justin Bacon, Prairie Grove FFA Chapter)*

Right: This unique barn sits in the middle of the countryside. Its main purpose is to house horses. The many different turns and odd shapes of its architectural style make it truly stand out. *(Submitted by Amanda Bacon, Prairie Grove FFA Chapter)*

This old barn may have a rusted roof, but its white elegance stands out on this curvy country road. This **Washington County** barn stands among houses now but was once in a very rural area. *(Submitted by Amanda Bacon, Prairie Grove FFA Chapter)*

Above: Daybreak Farm is located on **Highway 25B** in **Cleburne County**. There are four barns on the farm. This one, along with the windmill, was built in the 1960s and re-skinned in 2002. The Daybreak Farm is in the cattle business. *(Submitted by Hayden Henson, Quitman FFA Chapter)*

Left: This weathered clapboard barn outside of **Lincoln** in **Washington County** is used in the farmer's beef cattle operation. Like most beef farms in Northwest Arkansas, this farm also has a poultry operation. In fact, the barn sits near several chicken houses. *(Submitted by Cassie Bacon, Prairie Grove FFA Chapter)*

Right: In the **Viney Grove** community, near **Prairie Grove**, this old rock barn sits in the middle of a beef cattle field. Formerly used as a milking barn for dairy cows, it now sits vacant. *(Submitted by Amanda Bacon, Prairie Grove FFA Chapter)*

Below: The O. D. Gunn Trade & Sale Barn is located in downtown **Quitman** behind the Circle L Feed Store. This barn is almost one hundred years old and still in great shape. The barn played a huge role in the history and development of Quitman but is not in use today. *(Submitted by Hayden Henson, Quitman FFA Chapter)*

This old "Civil War" barn was built in the late 1800s. It is located between **Clinton** and **Fairfield Bay** in **Van Buren County** on **Burnt Bridge Road**. This classic weathered barn is still used today after 129 years of service. The lumber used to build the barn was acquired from some houses built during the Civil War (top). In January 1993, the barn showed wear and tear. It was still used at this time, however (above left). Amidst the fall foliage in September 1998, the damage to the roof is evident (above right). In August 1999, one can see the repair to the roof. This photo was taken just before the fall colors came in (left). *(Submitted by Ken Crocker)*

Left: Wiley's Cove Ranch barn is located between **Leslie** and **Harrison**. Unique in its elongated shape, it has been refurbished and is still used today. Photographer Ken Crocker first noticed the barn in October 1990 when traveling to Branson (top left). In May 2001, the barn was repainted. The October 2003 fall foliage enhances the beauty of the barn (bottom left). *(Submitted by Ken Crocker)*

Below: In 1997, during one of Ken Crocker's back road barn searches, he stumbled across this classic. Located in **Scotland**, just west of **Clinton**, it is used to store hay for the horses. *(Submitted by Ken Crocker)*

While visiting his folks in **Fairfield Bay** in January 1993, Ken Crocker snapped this photo of an old barn. The barn is located in **Van Buren County** overlooking **Burnt Bridge Road**. *(Submitted by Ken Crocker)*

The Hubbard family barn is located on **Highway 16**, east of **Clinton**. The barn was built in 1925. This photo was taken in October 1990. *(Submitted by Ken Crocker)*

In the foothills of the Ozarks, just west of **Fairfield Bay** and north of **Burnt Bridge Road**, the Fred Herman barn sits quietly. Herman still uses the barn, built in the 1940s. This photo was taken in January 1996. *(Submitted by Ken Crocker)*

Red River Riding Stables sits just outside the main entrance to **Fairfield Bay** in **Van Buren County**. Steve Swenson, the owner/operator of Red River Riding Stables, built this barn and displays some items of the past. This photo was taken in the spring of 2000. *(Submitted by Ken Crocker)*

In February 2003, I located this beautiful old hay barn just north of **Lamar** in **Johnson County**. *(Submitted by Ken Crocker)*

This old weathered barn has seen better days. Even ragged and worn out, it retains its majestic presence on the farm. The barn is located in **Oppelo, Conway County**. *(Submitted by Ken Crocker)*

This weathered red barn was photographed in June 2003. It is located in **Pearson** between **Heber Springs** and **Fairfield Bay**. The barn is still used today and is very well maintained. *(Submitted by Ken Crocker)*

This classic wooden barn is located in the foothills of the Ozarks in **Newton County**. The barn is nestled among the trees and is used for working cattle. The photo was taken in May 2001. *(Submitted by Ken Crocker)*

This wooden barn and corncrib are nestled in a grove of trees and enhanced by the beauty of the spring season. This barn is located just north of **Boxley** in **Newton County**. Photo taken in May 2001. *(Submitted by Ken Crocker)*

Weathered boards and a rusty tin roof showcase this classic barn located on **Highway 21** just north of **Boxley** in **Newton County**. This photo was taken in May 2001. *(Submitted by Ken Crocker)*

In 1989, this barn served as a display for First National Bank (above). The barn is located in **Van Buren County** just west of **Fairfield Bay**. In 2002, David Baron, the present owner, restored and repainted the barn. It is used to house his two camels that are regularly seen from Highway 16. *(Submitted by Ken Crocker)*

From the front or the side, this barn is a classic. It is located in the **Buffalo National River** area in the Ozarks. I found this barn in October 2003 on one of my many barn searches. *(Submitted by Ken Crocker)*

Deep in the **Buffalo National River** area, down a couple of mud roads, you can find this old farmstead, which was established in 1830. The buildings are well maintained and fabulous to look at. These photos were taken in October 2003 on a cold and rainy day. *(Submitted by Ken Crocker)*

While traveling down **Highway 21** just north of **Boxley**, there are many barns to see. This barn has a unique look that is enhanced by the trees that surround it. This photo was taken in May 2001. *(Submitted by Ken Crocker)*

Above: Classic features, weathered boards, and rusted tin showcase this wooden barn in **Newton County**. This barn is close to the **Buffalo River** area and is still used. Photo taken in May 2001. *(Submitted by Ken Crocker)*

Right: This old lean-to barn has seen many years of service. It is located in **Newton County** near **Boxley** and is surrounded by lush fall foliage. This photo was taken in October 2003. *(Submitted by Ken Crocker)*

Left: This barn is located in **Clinton, Van Buren County**. You can only see it by looking across the grassy runway at the Clinton Airport. The photo was taken in the summer of 2003. *(Submitted by Ken Crocker)*

Below: Norm Luke, a native of Shirley and the owner of the Barber Shop, sent me out to find this barn. It is located on the back roads up in the hills behind the Barber Shop, just west of **Shirley**. Although the barn has seen better days, it has a great deal of character. This photo was taken in October 1996. *(Submitted by Ken Crocker)*

Just south of **Boxley** on the way to **Harrison**, this elongated wooden barn is located in the foothills of the Ozarks in **Newton County**. This photo was taken at the height of the fall foliage season in October 2003. *(Submitted by Ken Crocker)*

Above: Norm Luke, a native of Shirley and the owner of the Barber Shop, told me this barn was over one hundred years old. You can find it on **Highway 16** as you drive into **Shirley**. The photo was taken in October 1996. *(Submitted by Ken Crocker)*

Left: This old wooden barn is located in **Harrison** just north of the cemetery. We discovered it while we were out taking fall foliage pictures. My brother-in-law said the barn was built in the 1930s. Photo taken in October 2003. *(Submitted by Ken Crocker)*

Right: Boone County and **Harrison** offer a barn lover many opportunities to find rare treasures. This barn is located on **West Branch Road**. It was raining the day I photographed it, but I never had to leave the truck. A gorgeous sight without the muddy feet I expected! *(Submitted by Ken Crocker)*

Below: West Branch Road in **Boone County** is barn-rich! Someone has been working on this old barn as evidenced by the ladders against the front. Other than a little roof damage, this barn is functional. Photo taken in October 2003. *(Submitted by Ken Crocker)*

In October 2003, my brother-in-law and I jumped in his four-wheel-drive truck and headed for the back roads of **Harrison**. We found this old barn, which has seen quite a bit of action. Although the barn may have had some repairs, it looks like it needs a little more fixing up. *(Submitted by Ken Crocker)*

Dr. Sam Scroggins owns this beautiful barn, which is located in **Boone County**. Its look is not "traditional" in any sense of the word, from the rounded flow of the roof to the interesting rock/shingle/wood siding. This barn has style. *(Submitted by Ken Crocker)*

I found this barn on my way to Fairfield Bay in June 2003. It is located in **Logan County** between **Fort Smith** and **Russellville**. It doubles as a storage unit; round bales surround it. *(Submitted by Ken Crocker)*

This 1994 photo depicts the Leland Parks shed in **Van Buren County**. In the 1890s, Zachary Parks built his single-crib barn of pine logs using the square-notching style. This barn is no longer standing. *(Submitted by Cindy Wilson, Van Buren County Farm Bureau Women's Committee)*

Located in **Cleburne County** on **Highway 25**. *(Submitted by Geneva Wallace, Wynne FFA Chapter)*

Above: Orville and Bernice Horton built this **Searcy County** barn in the 1930s. It is used to house livestock and store hay. *(Submitted by John Estes, Marshall FFA Chapter)*

Right: Preserving our past is important. W. H. Parks built this four-crib barn in 1912. It is constructed of oak logs using the saddle-notching style. This 1994 photo taken near **Dennard** documents a vanishing way of life. This barn is no longer standing. *(Submitted by Cindy Wilson, Van Buren County Farm Bureau Women's Committee)*

Arthur A. "Ott" and Virginia Nicholson, Uncle Bob Babe, and Oscar "Snooks" Reeves built this **Cleburne County** barn in 1965. Henry Reeves sawed the lumber. The Nicholson family used the barn for storing hay and raising cattle, horses, pigs, and every other farm animal. Now, Ott's son, Curtis A. Nicholson, owns the barn. Geneva Nicholson Wallace married her husband in the pasture in front of this barn. *(Submitted by Geneva Wallace, Wynne FFA Chapter)*

Right: Jim Dossett, Oscar "Snooks" Reeves, Gatis Sample, and Lum Nicholson built this red-roofed barn in the 1930s. Doyle Dossett, the original owner, used the barn for his white-faced cattle. Ruby Reeves now owns it. *(Submitted by Geneva Wallace, Wynne FFA Chapter)*

Below: Delbert Harris, a farmer who used the barn for his mules, three head of cattle, and hay, built this **Cleburne County** barn in 1950. His stepson, Tony Porter, now owns the barn. *(Submitted by Geneva Wallace, Wynne FFA Chapter)*

Above: Arthur A. "Ott" Nicholson and Oscar "Snooks" Reeves built this barn around 1968 for William and Mary Warbington. Jerry and Michelle Reesor of Floral now own the barn. The Reesors use the barn for hay storage. *(Submitted by Geneva Wallace, Wynne FFA Chapter)*

Right: How old is this barn? This **Washington County** barn shows much wear and use. But it still keeps the hay out of the weather and stores odds and ends for this diverse farm. The owner has cattle, goats, mules, and sheep. *(Submitted by Cassie Bacon, Prairie Grove FFA Chapter)*

Located in **Marshall**. *(Submitted by Clyde Dawes, Advisor, Mountain View FFA Chapter)*

Top three photos: As you enter the Gilson property, your view is swept up a curving drive into the two-story red barn. This barn was converted seven years ago and put to good use. The two garage doors provide an entrance into the lower level where Mr. Gilson stores chemicals for his business. The loft has been converted to a home for one of his workers.

From the main house, the whole barn is visible. The loft is accessed through a balcony stairwell. Mr. Gilson's yellow lab often dozes on the roof.

The back of the barn offers a view of the plant patio and the entrance to the loft. A garage door allows trailers to be pulled through. This enables Mr. Gilson easy access to his chemicals. *(Submitted by Angela B. Collins, Arkansas FFA State Executive Secretary)*

This bright red barn is found on one of those crooked little roads in the country. Or is it in the country? This barn is actually just south and west of **Little Rock**. It is near **Alexander**, and the Walshes own it. An older structure peeks out from behind. *(Submitted by Angela B. Collins, Arkansas FFA State Executive Secretary)*

The Harris barn stands proudly next to **Interstate 40** in **Pulaski County**. It is used for storage of round hay bales. This barn has seen its share of wind damage but will stand the test of time. *(Submitted by Eva Fletcher, Director of Camp Couchdale)*

Above: Nestled in the woods, this barn hasn't been touched for years. Built around fifty years ago, it still maintains its majestic beauty. *(Submitted by Renee Martin, Hamburg FFA Chapter)*

Right: The H & S (Hal and Susan Gibson) mule barn, located in Selma, was built in 1992. Carrie, Hal and Susan's daughter, posed for a picture while the barn was under construction. The barn is now completed and boasts eight twelve-by-twelve-foot stables, a feed crib, a tack room, and a loft. Hal and Susan raise mules. *(Submitted by Hal Gibson, Advisor, Hamburg FFA Chapter)*

Above: "Floyd's Barn" was located in a lot on the Gibson Brothers Farm in **Selma, Drew County**. It had an equipment shed on the left; a corncrib and cottonseed hull crib in the middle; and a stable on the right. Glynn Gibson is seated on Colonel, next to an old Oliver cultivator. *(Submitted by Hal Gibson, Advisor, Hamburg FFA Chapter)*

Right: This delta land barn is located on a canal bank between **Arkansas City** and **McGehee**. The barn is used for livestock (hayloft and hay mangers attest to that) and is flanked by a larger barn to its rear. *(Submitted by Hal Gibson, Advisor, Hamburg FFA Chapter)*

This barn calls you to a time in the past. Its classic style and setting invite you to sit in the shade and sip lemonade. Mr. Brown built it in 1964. The old rail line used to run from Benton to Hot Springs. When the line was torn up, Mr. Brown acquired the lumber and built this barn. It's not very often you see a barn built of railroad trestles! Mr. Brown stores an old-fashioned buggy on one side and remembers the days of low gas prices when he gazes at his old-fashioned gas pump. Mrs. Brown keeps the grounds beautifully landscaped. The overall picture is perfect. *(Submitted by Angela B. Collins, Arkansas FFA State Executive Secretary)*

This old barn is located just off **Interstate 30** near **Bryant**. It overlooks the interstate, and railroad containers surrounded it. Even overgrown with weeds, this barn grabs your eye. *(Submitted by Angela B. Collins, Arkansas FFA State Executive Secretary)*

This page and opposite: The "Bayou Barn," located in **Drew County** on **Bayou Bartholomew**, once served as a home to mules that worked the cotton fields. The corridor down the center (left) was used to walk the teams to the bayou for water. To the left of that corridor, there are three large feed cribs and a tack room. To the right of the corridor (opposite at top), there were double feed stalls and hay mangers for the teams. Hay was pitched down through the holes in the loft (opposite at bottom) to feed the mules. In later years, the barn was used to raise registered Hereford show cattle. *(Submitted by Hal Gibson, Advisor, Hamburg FFA Chapter)*

Al and Marie are retired from the federal government. When they left Washington, D.C., they decided to head home. Marie is originally from Saline County, but they had been gone for forty-two years. They found an old farmstead and decided to settle there. The old farm included two homes and several old barns on ten acres. The ninety-four-year-old daughter of the original farm owner still resides on site.

In 1998, Al and Marie started construction on a new barn to house Al's cars and hobbies and then proceeded to build the "Great House." An effort was made to restore/retain the old structures; most are still intact and in use. The rabbit house was one of their first attempts at reconstruction. The first few yanks on the boards, however, resulted in complete destruction. A new rabbit house stands today and houses the yard equipment.

Another project involved the pond. When Al and Marie bought the farm, a huge pond dominated most of the property. The culprits: beavers. After tearing out the dam, they drained much of the land and brought the pond back to its former size. Several gardens dot the landscape. One garden of particular interest is located where the beavers erected their dam. That garden is affectionately referred to as that "Dam Garden."

Al and Marie have lived here for six years. They love to work around the place, picking up pecans, cultivating their garden, and entertaining visitors. *(Submitted by Angela B. Collins, Arkansas FFA State Executive Secretary)*

The Smoke Shack is located in the **Bethel** community on the outskirts of **Glenwood**. James Bardwell built it in the late 1800s, despite having lost an arm. Also called the "Sugar Shack," the floor is dirt with a hole dug in the middle. A fire would have been built in the hole to smoke meat. Until the 1950s, the shack was used to smoke pork and to store kraut. The Smoke Shack still stands today. *(Submitted by Jessica Miller, Centerpoint FFA Chapter)*

Right: During the February 2002 storm, several cows sought shelter in the Potter barn. Located near **Nashville** in **Howard County**, this barn was built in the late 1930s. The Woodruff family bought the barn in the 1950s and continues to use it today. *(Submitted by Ben Wallis, Nashville FFA Chapter)*

Below: South Arkansas was blanketed in snow in February 2002. This old barn, located in the **Center** community, near **Nashville**, was built in the 1920s. It is situated near an old house where the Wallis' great-grandparents lived as newlyweds. *(Submitted by Ben Wallis, Nashville FFA Chapter)*

The Bar 50 Ranch is a bed-and-breakfast in **Hot Spring County**. The facility is owned and operated by Bette Clay and offers full lodging for riders and horses. The barn has eighteen stalls and numerous turnout paddocks. The B & B is open year-round; has facilities for company meetings, family reunions, and vacationing; and offers a petting zoo and rental horses. *(Submitted by Nina Crutchfield, Advisor, Searcy FFA Chapter)*

This barn, built in the 1930s, is located on the A. J. and Chester Woodruff farm in the **Center** community. Originally used as a barn, it was converted to a peach shed in the early 1950s. In the early 1990s, it was converted back into a barn and is still in use today. *(Submitted by Deb Wallis, Nashville FFA Chapter)*

This **Pike County** barn was built in the 1950s. It is located on the old Strawberry Road just north of **Delight**. *(Submitted by Megan Antley, Hope FFA Chapter)*

On one of my back road barn searches, I found this red beauty just east of **Royal**. The classic barn shows years of use, along with wear and tear from the elements. This photo was taken in June 2003. *(Submitted by Ken Crocker)*

In 1999, a series of tornados swept through **Howard County**, leveling many homes and barns. Chester Woodruff was one of the farmers affected. This new barn was built in 2001 to replace one of Woodruff's lost barns. *(Submitted by Deb Wallis, Nashville FFA Chapter)*

Above: On a return trip to Dallas, I spied this barn from Interstate 30. It is located just east of **Hope** in **Hempstead County**. The photo was taken in October 1989. *(Submitted by Ken Crocker)*

Left: This barn is located on the Hopson family farm. It was on site when they moved there in 1940. The lumber for the barn was cut from logs on the place. It has two corncribs, storage for hay, and animal stalls. The family enjoys having this piece of history on their homeplace. *(Submitted by Hannah Wilhite, Acorn FFA Chapter)*

Nestled under the oak trees in **Howard County**, this old barn tells quite a story. It is the story of time and the elements. Although settling and showing its age, the barn is still used for shelter and storage. This photo was taken in June 2003. *(Submitted by Ken Crocker)*

This lovely little horse barn is located just three miles from Arkansas FFA Camp Couchdale in **Garland County**. The owners are retired and operate this small acreage farm just off **Highway 290**. *(Submitted by Eva Fletcher, Director of Camp Couchdale)*

Mr. A. C. Hale, the first agriculture instructor at Fairview High School, and his students built this barn around 1940. Several barns in the Gothic style were built in the surrounding area. The excessive height allowed for hay storage. Many farmers during this time period used loose hay rather than baled hay. Hay would be hauled to the barn in a horse-drawn wagon and unloaded with pitchforks through an opening in the rear of the barn. The ground level of the barn is divided into four quarters: chicken house, stanchions for milking cows, horse stalls and feed storage, and a holding area for stock. *(Photo submitted by Daniel Curry, Camden Fairview FFA Chapter. Information provided by Mr. Ralph Hale.)*

This is Dr. Randy Burgess's Ouachita Equine Clinic. This clinic/barn houses state-of-the-art equipment for the care and recovery of horses. It has operating rooms, stalls, and everything needed to house horses safely. Dr. Burgess is a great supporter of the Arkansas FFA. *(Submitted by Hannah Wilhite, Acorn FFA Chapter)*

Above: These two weathered beauties are located on **Highway 371**, north of **Laneburg** and south of **Prescott**. They are used to store hay and to provide shelter for the horses and cattle in the foreground. *(Submitted by Eva Fletcher, Director of Camp Couchdale)*

Right: Humphrey's dairy farm is located in **Garland County** on **Shady Grove Road**. The dairy is still in operation and provides daily milk deliveries to customers. A fire in one end of the dairy barn destroyed the roof; it has been replaced. This barn is constructed of rock and houses a Holstein dairy herd. *(Submitted by Eva Fletcher, Director of Camp Couchdale)*

This barn is located on **Highway 290**, about six miles from Arkansas FFA Camp Couchdale. It is constructed of logs and may have been a one-room cabin in its early years. *(Submitted by Eva Fletcher, Director of Camp Couchdale)*

Driving to Camp Couchdale is always a chance to get back to nature. You can see the mountains, timber, and this barn. Located on **Highway 171**, this barn belongs to Loyd and Mildred Smith. Their cattle love to hang out around this barn and the big tree; at times, the cattle dot the landscape. This is only one of the many barns on the Smith place. *(Submitted by Angela B. Collins, Arkansas FFA State Executive Secretary)*

Two bottom photos: When Renee and the late J. O. Lucy bought their home, they discovered that the barn and the house were part of the original Warrick homestead. Several other homes were built in the same area by the rest of the family. The barn is over sixty years old, and Renee uses it for her horses. She has two riding horses—one Arabian and one quarter horse. In the summertime, the horses retreat to the center of the barn, which is always ten degrees cooler than outside. Interesting fact: there are no posts in the loft holding the roof up. The roof is cantilevered! All the support is pushed to the outside of the structure. The foundation of the barn is rock with quite a bit of quartz showing. The Warricks used the barn for their draft horses, cattle, feed, and tack. *(Submitted by Angela B. Collins, Arkansas FFA State Executive Secretary)*

Papa Check's barn was built in 1967 in **Howard County** near **Nashville**. This photo was taken in February 1985 just after a good snow. His five grandsons gave Papa Check his name; others know him as Chester. In 1999, Papa Check lost this fabulous barn to a tornado. *(Submitted by Deb Wallis, Nashville FFA Chapter)*

Index of Photo Contributors

About the Author

Angela B. Collins is the Arkansas FFA State Executive Secretary and an agriculture education program advisor. Her duties include training the Arkansas FFA State Officers, providing leadership programs throughout the summer, serving agriculture teachers, and planning and running the Arkansas FFA State Convention. One thing is certain: there are never any dull moments in her work. She has served in this capacity since January 2000. Prior to her service in the Department of Workforce Education, she taught agriculture and biology in the classroom. She taught in Wynne, Arkansas; Covington, Louisiana; and Ponchatoula, Louisiana.

Angela enjoys working with FFA members throughout the state. When she is not working with FFA members or agriculture teachers, she enjoys reading and working in the yard.

A native of Louisiana, Angela graduated in 1992 from Louisiana State University. She and her husband, Charles "Coo" Collins (also an LSU graduate), moved to Arkansas in 1994. They currently live in Bauxite, Arkansas, with their daughter, Darcey. They are expecting their second child in March 2005.